The Karakul Sheep in America

also known as Persian Lambs or Qaraqul Lambs

by J. Walter Jones

with an introduction by Jackson Chambers

Self Reliance Books

Get more historic titles on animal and stock breeding, gardening and old fashioned skills by visiting us at:

http://selfreliancebooks.blogspot.com/

Introduction

I am pleased to present yet another practical title on breeding and raising livestock.

The work is in the Public Domain and is re-printed here in accordance with Federal Laws.

As with all reprinted books of this age that are intended to perfectly reproduce the original edition, considerable pains and effort had to be undertaken to correct fading and sometimes outright damage to existing proofs of this title. At times, this task is quite monumental, requiring an almost total "rebuilding" of some pages from digital proofs of multiple copies. Despite this, imperfections still sometimes exist in the final proof and may detract from the visual appearance of the text.

I hope you enjoy reading this book as much as I enjoyed making it available to readers again.

Jackson Chambers

The Celebrated Blackfaced Ram "CAIRNTABLE,"
Bred by CHARLES HOWATSON, of Dornel, and sold for SIXTY POUNDS,
To J. N. FLEMING, of Keil   July, 1870

THE
KARAKÚL SHEEP
in America

Teddy, Sr. the best Karakul ram of the first importation.
Now to be seen on Bunbury Farm, Charlottetown, Prince Edward Island.
Property of the Dr. C. C. Young Karakul Sheep Co. Limited.

The Karakúl or Arabi Breeds
of Sheep.

Five years ago there were no Karakul sheep in America. During the past five years two importations of these breeds of sheep were made by Dr. C. C. Young. The first lot, consisting of five rams and ten ewes, was purchased in 1908, and the second importation consisting of 11 rams and 6 ewes arrived in quarantine at Baltimore in March, 1913. In the first importation only two unrelated rams proved to be valuable; viz Teddy, Sr. and Fasset, but because of the experience gained and discoveries made by Dr. Young, he obtained at least six distinct, unrelated, blood lines of high quality in the second importation. One of the valuable rams died in quarantine, thus reducing the total number of distinct blood lines on the American continent to not less than seven unrelated rams—all of which are now alive and in good health. No other Karakul sheep have been imported to America. **Importa-tions**

Blood Lines

Six of the seven strains are now owned by the Dr. C. C. Young Karakul Sheep Company, Limited, of Charlottetown, P.E.I., Canada. They are: Teddy, Sr., Fasset, Vaska, Nezaekiev, Poltava, and Baron von der Launetz. The first five were placed in the breeding pens with 400 Lincoln, Highland Black Faced, Karakules, Leicester, and other long-wooled ewes at three farms near Charlottetown; viz Upton, Bunbury, and Dinnis; the sixth ram was loaned to the United States Department of Agriculture at Washington and arrived on Prince Edward Island in January, 1914, and is now being bred to forty Lincoln ewes. The only other unrelated Karakul blood in America is on a New Mexico ranch, and Dr. Young owns a half interest in all the full-blood Karakules in that flock. **Location of the Industry**

The home of the Karakule breeds of sheep is the Khanate of Bokhara—a Russian quasi-dependency for some past fifty years. It is situated in West Turkestan, of Central Asia and its capital—Old Bokhara city— in which the Emir resides, is about 600 miles east of the Caspian Sea and 200 miles north of Afghanistan. The **Home of the Karakul Sheep** natives are uncivilized, profess the Mohammedan religion, and, until conquered by the Russians, were desperate desert robbers and fighters. They are to-day closely guarded by Russian Cossacks located at military posts throughout the entire West Turkestan. Foreigners are permitted to travel only in well-defined zones on special permits issued by the war minister of Russia, and cannot enter the ulterior districts where the best Persian Lamb producing areas are. The achievements of Dr. Young

No. 2. Vaska—one of the best rams of the second importation—being held by the author of this book on his farm "Bunbury" near Charlottetown P.E.I., Canada. In the background is Hutchinson Harris, the well known fur merchant of London, Mr. F. B. McRae, and Dr. C. C. Young.

in obtaining herds from forbidden territory will be fully described in a book he is preparing on Bokhara and the Karakul Sheep Industry. It is only necessary to state in this booklet that we believe it is now impossible for foreigners to obtain any more sheep out of Bokhara. This fact has been well proved by correspondence with Russian authorities who quote the laws of Bokhara and Russia and is further certified by Mr. Karpov in a bulletin issued by the Russian Imperial Department of Agriculture in which Dr. Young's method of obtaining his herds is criticized and amendments to the existing laws are recommended to prevent a recurrence of importations. It

No more Importations

is not probable that even if the strongest pressure and influence were used more unrelated Karakul rams could be obtained for America from Bokhara. And even if they were obtained the veterinary authorities would not permit them to be brought into America.

No. 3. Dr. C. C. Young in the Circassian National Costume at Baku, Asia Minor. The Bourka robe about him is woven from camel's hair.

The Directors of the Dr. C. C. Young Karakul Sheep Company believe that they hold a monopoly of the industry on the American Continent by controlling the blood lines. The fatal effects of inbreeding on sheep are well known to scientific sheep raisers and no other breeders of Karakul sheep in America can continue without close inbreeding unless they purchase rams from the Dr. C. C. Young Karakul Sheep Company. Inbreeding will largely destroy the character of the fur by making the fibre too fine, thus loosening the curl in the lamb skin. It is proposed by the Directors to control the produce of the parent Karakule sheep company by retaining a

Controlling the Blood-lines

Evil Effects of Inbreeding

7

majority control on all rams sold and by forming subsidiary companies whenever rams are available in sufficient numbers. It is thought that at least a dozen subsidiary companies can be formed in 1914 as already six mature rams of proved quality are available for that purpose.

When Karakuls are crossed with native longwool-breeds such as the Lincoln, Highland Black Face, Leicester,

No. 4. Afghan fine-wool sheep (white)
held by Sarts of Bokhara at Tjar-jui. The great majority of sheep in Bokhara are of this strain, the fine wool of which is reducing the grade of persian lamb fur year by year. These sheep show traces of Karakul blood.

Karakuls Crossed with Native Longwools Cotswold, or any sheep with coarse wool fibre with no fine underwool intermixed, the progeny is a black, lustrous, tightly-curled lamb which when killed and skinned is worth from six to thirteen dollars in wholesale lots. The half-blood rams if reared to maturity will frequently produce as good results in the quarter-blood lamb. The Middlewater Cattle Company who formerly owned some of the herd now at Charlottetown produced a large number of skins of an average value of $6.79 each. even though the great majority of their ewes were not free from Merino fine-wool blood. It is not claimed that the Charlottetown herds of native ewes are free from fine wool, but as fine a selection as possible of Lincolns was made and a liberal percentage of ewes will throw lambs whose skins will obtain the highest prices obtainable for Persian Lamb fur. Inasmuch as there are no lustrous-wooled sheep in Asia like our English Longwools and Karakul breeders are compelled to cross with Koordink (fat-rump) and Afghan fine-wool strains it is evident that fine-wool has been introduced into almost all the Asiatic herds and the industry been nearly ruined by unscientific mating. It is fully expected that the elimination of fine wool, and the increased lustre of the Maritime Provinces sheep, together with scientific mating will produce skins which will far excel those produced in Bokhara. Skins have already been produced from our rams even in the first cross which government furriers have certified are the equal of the best produced anywhere. **Evil Effects of fine Wool Admixture**

In a circular letter dealing with the Karakule Sheep Industry lately issued by the United States Department of Agriculture, it is stated that the total wholesale trade in Persian Lamb skins, which are produced only by Karakule Sheep, is $14,000,000.00 annually for the United States alone. A tremendous capitalization has lately been made of the newly domesticated silver fox, but the product of this rare and beautiful animal sells for less than one million dollars annually in the whole world. As a fur proposition the Karakul sheep would easily surpass the fox because the sheep is a fully domesticated animal and one male will mate 100 females (as has been proved), its wool and its mutton is better than those of ordinary sheep, while the lamb at birth is several times more valuable than the lamb of the common breeds at six months of age. Besides, the usual 20 or 30 per cent of still-born lambs (slunks) possess valuable fur so that they are not disposed of on the dung-heap, but fetch often as high as twenty dollars each. It is not proposed to cater to the production of fur for many years to come but to produce breeding animals to head and establish flocks all over the continent which will produce fur. The price of Persian Lamb fur has advanced, according to Brass, over 300 per cent in 20 years. **Volume of Annual Trade**

Comparison with Silver Fox for Profits

Parent Company's Control

There is another great advantage over ordinary sheep raising that alone would provide enough increase of value in the products to insure a great demand for breeding animals. It is the fact that when the ewes are bred early (in August) they produce lambs in January. **Breed Twice a Year**

Twins

These can be killed immediately and sixty per cent of the ewes will again produce lambs in July. If only single lambs, and not twins, as frequently occurs, were produced, it is possible that one ewe might produce $20 or $25 worth of fur yearly, or $40 to $50 if twins came. That being the case, what price would breeders pay for Karakul rams of good quality?

No. 5. A two days old Karakul lamb ready for killing.
The skin of this specimen is worth $12.00 wholesale value. The curls extend to the nose and hoof and around the belly, and are close or tight, wavy, and lustrous.

Karakuls are Hardy

Better Rustlers

Moreover, the Karakul sheep are much hardier than native breeds. This has been absolutely proved on ranges in the Panhandle of Texas where 60 per cent of Shropshires perished in a blizzard while not one Karakul succumbed. Personal observation by the directors during winter conditions on P. E. I. has confirmed them in the opinion of their great hardiness, compared with ordinary sheep. Their rustling qualities are the equal of goats.

10

The company, in order to reap the full benefits of **Absolute** their faith and enterprize, will be obliged to maintain **Control** control of their blood supply just as Silver Fox breeders once held control of the breeding stock of foxes. In the latter case they were finally beaten because of the fact that wild silver foxes were captured in the wilds by others. But, in the case of Karakul sheep, only importations from Bokhara can disrupt the American control. It is therefore prorosed and planned to maintain majority ownership af all fnll blood live stock sold until the business

No. 6. Four Rams of the First Importation.
Two of these rams—Lowden and Yermoloff died in 1912. The other two,— Teddy. Sr., and Fasset, together with Lowney who is not in the picture, are now at Charlottetown, P.E.I.

is well established in America. After that time the parent company will still control the sale of breeding rams. It is not anticipated that others can secure any importations from Bokhara.

There are at Charlottetown, P. E. I., the property **Live** of the Dr. C. C. Young Karakule Sheep Company, **Stock** the seven unrelated male blood lines mentioned above **now** and their descendants numbering nineteen full-blood **owned** rams and twenty-eight full-blood ewes. Only one of the rams and three of the ewes are fine wools. Fourteen of the ewes have been bred to Teddy Sr. in order to secure a considerable number of his progeny before he is too old for service.

Besides the full bloods there are at Charlottetown four choice, coarse-wool, Karakul—Persian fat-rump ewes, as well as ten Highland Black-faced ewes, and about three hundred and fifty ewes, pure-bred Lincolns, or crosses of Lincolns, Cotswolds and Leicesters,—making a total of upwards of four hundred sheep.

Assured Monopoly of Coarse Wool Blood Lines

The Karakul Sheep were secured from Dr. C. C. Young and the Middlewater Cattle Company, of Texas, whose herds were purchased outright. The only herd in America besides these, which deserves consideration as producers of persian lamb fur, is that in New Mexico and Dr. Young owns a half interest in it. What few other Karakul and grade Karakul Sheep there are in America, are wholly the progeny of the three fine-wool rams of the first importation, namely, Tawney, Yermoloff and Louden. The Crawford ranch in Kansas secured a grandson of Teddy, Sr. off Dr. C. C. Young's ranch in the fall of 1912 while he was absent in Asia, but have no other unrelated coarse wool blood free from fine wool among their sires. In 1913 they produced some persian lamb skins from this ram crossed with their fine-wool ewes.

Evil Effects of Inbreeding

Rearing Astrakhan from Fine Wool Karakuls

Unless unrelated strains of Karakul blood are available inbreeding will "breed the wool of the sheep's back" and make the wool fibre so fine that only Astrakhan fur can be produced. The Dr. C. C. Young Karakul Sheep Company, Ltd. is the only company that can prevent inbreeding. This point is of great importance as the wholesale price of persian lamb skins is as high as fifteen dollars in bale lots of 100 to 200 skins each, while Astrakhans rarely fetch over three and a half dollars each wholesale. The production of Astrakhan fur would hardly be more profitable than rearing for meat production, unless a proportion of the lambs could be classed as low grade persian lamb fur.

HISTORY AND PRESENT LOCATION OF THE KARAKUL SHEEP OF THE FIRST AND THE SECOND IMPORTATIONS JANUARY, 1914.

First Importation	Animals	Wool Characteristics	Present Location	Remarks
Imported 1908, producing first lambs in 1909.	*Teddy, Sr.	Coarse, no fine	Charlottetown, P.E.I.	Purchased from Middlewater Cattle Co.
	*Fasset	Coarse, no fine	Charlottetown, P.E.I.	Purchased from Middlewater Cattle Co.
	Tawney	Fine wool	Charlottetown, P.E.I.	Purchased from Middlewater Cattle Co.
	Yermoloff	Fine wool	Dead	Lowden Ranch, Texas in 1912.
	Lowden	Fine wool	Dead	Dr. C. C. Young Ranch, Texas, in 1912.
	5 Ewes	{ 2 Coarse wool	Charlottetown, P.E.I.	Purchased from Middlewater Cattle Co.
	5 Ewes	{ 8 fine wool	{ Kansas, one dead.	4 purchased by Crawford Ranch from Dr. C. C. Young's ranch, Coahuila, Mex.
Second Importation				
Imported 1913; a few lambs were born in quarantine, which were of pure blood, but their parentage is unrecorded.	Unnamed Ram.	Coarse wool	Dead	Succumbed in quarantine at Baltimore.
	*Vaska	Coarse wool	Charlottetown, P.E.I.	Purchased from Middlewater Cattle Co.
	*Nezakiev	Coarse wool	Charlottetown, P.E.I.	Purchased from Middlewater Cattle Co.
	*Poltava	Coarse wool	Charlottetown, P.E.I.	Purchased from Middlewater Cattle Co.
	*Baron von der Launetz	Coarse wool	Charlottetown, P.E.I.	Used in 1913 by U.S. Department of Agriculture, Beltsville, Md.
	*6 unnamed Rams	Coarse wools	New Mexico	Possibly only one blood line as they were purchased off one ranch by Dr. C. C. Young.
	4 Ewes.	Coarse wool	New Mexico	Purchased from Dr. C. C. Young.
	2 Ewes	Coarse wool	Charlottetown, P.E.I.	Purchased from Middlewater Cattle Co.
	2 Ewes (lambs)	Coarse wool	Charlottetown, P.E.I.	Born in quarantine at Baltimore.

*The seven starred rams are considered to be absolutely unrelated blood lines and represent the only coarse wool Karakul blood on the American Continent.

It is to be noted that six of the seven blood lines are the property of the Dr. C. C. Young Karakul Sheep Co., Limited, Charlottetown, P.E.I.

Facts concerning the Karakúl Sheep Industry.

1. *First Importation to America, 1908, by Dr. C. C. Young.*

2. *Second Importation to America, 1913 by Dr. C. C. Young.*

3. *No other Karakul, or fur-bearing sheep, ever imported to America.*

4. *At least seven distinct, unrelated, blood lines of Karakul coarse-wool sheep in America.*

5. *Six of the seven unrelated blood lines are owned by the Dr. C. C. Young Karakul Sheep Company, Charlottetown, P. E. I.*

6. *Only coarse-wool sheep, free from an admixture of fine underwool, produce the highest grade of lamb skins.*

7. *Inbreeding of sheep is fatal to coarse wool. If the wool becomes fine, the curl in the lamb is more open and hence, less valuable.*

8. *The Dr. C. C. Young Karakul Sheep Company is the only owner of Karakul Sheep in America which can maintain a flock without inbreeding.*

9. *No more sheep of the required quality can be brought out of Asia into America under the present regulations in (1) Bokhara (2) In Russia (3) In United States and Canada.*

10. *The lambs produced by coarse-wool Karakul rams X coarse wool ewes of any breed and by half-Karakul rams X coarse wool ewes of any breed and often by quarter-Karakul rams X coarse wool ewes will average twice as high in value at birth as domestic lambs will fetch for meat at 6 months of age.*

11. *Lambs born dead, or aborted, are as valuable for fur as those born healthy.*

12. *A proportion of the ewes, about 60 per cent, can be bred twice a year.*

14

Facts concerning Persian Lamb, Astrakhan, Broadtail, and Krimmer Fur,

which is the skin of the lamb of the Karakul or grade-Karakul sheep.

1. *PERSIAN LAMB fur is the dressed and dyed skins of young Karakul lambs, or grade-Karakul lambs. The lambs may possess only 25 per cent of Karakul blood and yet grade as "persian lamb" but the ancestors must be coarse-wools (of any breed) and possess little, or none, fine-wool blood in order to produce tight curls,*

2. *ASTRAKHAN fur is the dressed and dyed skins of young fine-wool Karakul lambs or grade-Karakul fine-wool lambs which do not possess regular tight-curl formation. In almost every instance the Astrakhan fur is produced because of the presence of fine-wool blood in the ancestry of the lamb.*

3. *BROADTAIL or BABY LAMB fur is the lamb skin of the coarse-wool Karakul or grade-Karakul, aborted sometime before the close of the regular gestation period, which is five months. The abortion is not brought about artificially, as commonly supposed.*

4. *KRIMMER is grey lamb with curls similar to those of persian lamb or open curled like astrakhan. It can be produced by Karakul × white coarse wool sheep with about one eighth of Karakul blood. It is dressed without dyeing.*

5. *The comparative wholesale Prices of the Best Skins are as follow:*

Broadtail	$25
Persian Lamb	15
Astrakhan	4

6. *The comparative wholesale prices of average skins are as follow:*

Broadtail	$12.00
Persian Lamb	9.00
Astrakhan	3.00

    Retail prices are two or three times as high.

7. *Persians skins have advanced 300 per cent in price in the last 20 years.*

15

No. 8. Gate of the Palace of the Emir of Bokhara-Seid Alim Khan.

PHOTO BY Dr. C. C. YOUNG.

No. 9. On the left stands Atachadja Mirachur, Custom House Inspector for
Bokhara and the chief assistant to his Highness the Kushbegi; on the right
is Chief of Police of Bokhara.

SOME DIFFICULTIES ENCOUNTERED IN
IMPORTING KARAKULS.

Dr. Young describes the difficulties to be encountered by foreigners who seek to secure the Karakul sheep, describes his own peculiarly favorable position to perform the work, and ascribes to Russian courtesy his success in having secured sheep for America.

———

Passports and Purchase of Animals

Providing a RUSSIAN SUBJECT can secure permission from the Russian Department of Agriculture, and providing also that the Department of Foreign Affairs gives its consent, and providing His Majesty the Emir of Bokhara permits him to enter the Khanate, it is only a matter of being properly financed and interpreted and a limited number of Karakul sheep can be obtained, although, after all, the proper jigit, who is an official of the district governor and who can do what he pleases with his subjects, is indispensable in effecting the purchase.

Russian Restrictions

A FOREIGNER can not get out Karakul sheep for the following reasons: (1) Even after securing permission from the Russian War Minister to enter West Turkestan—which he will never get without the most active support of his Ambassador—he is not permitted to go to Turkestan of Central Asia at all, and he can absolutely not go to Takta Bazar, Kushk, Kerki, Termez or Karshi, which are in the forbidden military zone, and where valuable sheep may be found.

(2) The Emir does not permit a foreigner to export Karakul sheep and should he get them into European Russia through a third party, he can not get them out lawfully.

Quarantine Regulations

(3) Most all European countries prohibit the importation of live stock from Asia on account of certain diseases, and especially is that true of England, United States and Canada, where it is nearly impossible to secure a permit to land them; and even where an exception is made for purely experimental or exhibition reasons the most rigid quarantine is imposed, lasting for months.

Inborn Courtesy of Russians

Those foreigners who have traveled in Russia with proper credentials showing them to be interested in scientific research work will testify to the great courtesies shown them by Russian officials and this explains why I was able to get out a few head of Karakul, and on account of the mistakes made by me in my first and partly in my second importation I found after several months' investigation of the sources whence come our sheep, that some might possibly be inbred.   I hope to secure the permission to export a few more sheep and select them myself in the forbidden zone of Bokhara.   Though I was not permitted to enter the forbidden zone of Bokhara and Transcaspia last March, I hope to be able to do so this time, as it is the opinion of the Russian Department of Justice that since I became naturalized in America without the permission of the Russian Government, I am still a RUSSIAN CITIZEN, and need but return with a Russian passport, which gives the legal right to enter any section of the forbidden territory.

Dr. Young still "Russian"

DIFFICULTY IN SECURING PASSPORTS.

When Dr. Young sought permission from the British and Russian authorities to visit certain sections of the forbidden military zone of West Turkestan and a small northern section of Afghanistan, he found it impossible to secure the permission of either country to enter their territories. The following letter furnishes a proof of the effort made by the late United States ambassador to secure England's permission for Afghanistan.

American Embassy, London.

November 25th, 1912.

Dear Sir,—

Britain's Inability to Help

Your note of November 15th is at hand.

I regret to say that it is impossible to arrange with the British Government so unusual a privilege in a troubled borderland without instructions from the State Departments....If anything could be done in the direction you desire, it could better be done through our embassy at St. Petersburg; but I am under the impression that you would encounter the same difficulties there.....

Yours respectfully,

(Signed) WHITELAW REID.

Dr. C. C. Young,
    Grand Hotel d'Europe,
      Rue Michel,
        St. Petersburg, Russia.

———

The following extracts from a letter of the United States ambassador in St. Petersburg represents the result of the effort to secure permission from the Russian government to enter the desired district.

St. Petersburg, January 13, 1913.

Doctor C. C. Young,
    c/o Russian Diplomatic Agent,
        New Bokhara.

United States Inability to Help

My dear Dr. Young,—

If you will remember, your admission into the military district depended on your Russian friend who was to procure you privileges denied to other foreigners... Of course, as ambassador, I have been unable to secure any favours for an American citizen that are denied to citizens and subjects of other countries.......

Faithfully yours,

(Signed) CURTIS GUILD.

Prohibition of Importation

OF

Horses, Asses, Mules, Sheep, Goats and Swine from Asia and Africa.

Canada has similar regulations.

Ruling of the United States Department of Agriculture

United States Quarantine

Notice is hereby given to the owners, officers and agents of all steamers and other vessels of all descriptions plying between the countries of Asia and Africa and any of the ports of the United States, and Territories, or dependencies thereof, and to all stockmen and other persons concerned in any manner in the traffic in animals in or with the said countries of Asia and Africa, that certain contagious, infectious and communicable diseases, dangerous to the live stock of the United States exist among the animals of the said countries of Asia and Africa, viz. surra, affecting horses, mules and asses; foot and mouth disease, affecting horses, sheep, goats and swine; rinderpest affecting sheep, goats and swine.

Now, therefore, under the authority conferred upon me by the act of Congress approved February 2, 1903, entitled, "An act to enable the Secretary of Agriculture to more effectually suppress and prevent the spread of contagious and infectious diseases of live stock, and for other purposes" I do hereby prohibit the landing at any of the ports of the United States and Territories, or dependencies thereof, of any horses, asses, mules, sheep, goats and swine from the said countries of Asia and Africa. This prohibition shall take effect immediately and shall continue in force until otherwise ordered.

Done at Washington this sixteenth day of November, 1910.

(Signed) WILLET M. HAYS,
Acting Secretary of Agriculture.

A LETTER OF INTRODUCTION

from the Secretary of Agriculture of the United States.

THE UNITED STATES OF AMERICA

DEPARTMENT OF AGRICULTURE

To all who shall see these presents, greeting:

Be it known that Dr. C. C. Young, a citizen of the State of Texas, and of the United States of America, whose signature appears on the margin hereof, will in the near future, visit the various countries of Europe and Asia, for the purpose of securing such information and facts as may be obtainable relative to the character and breeding of sheep in those countries.

He is hereby introduced to all persons interested in this subject, whom he may meet.

(Signed and sealed September 13th, 1912)

————

A CIRCULAR LETTER

Issued in the fall of 1913 by experts of the United States Department of Agriculture on Karakul or Arabi Sheep.

KARAKUL OR ARABI SHEEP

The numerous inquiries directed to the Department of Agriculture concerning the persian lamb industry have led to the compilation of the following information. **Karakul Breeds the only Fur Producers**

Persian lamb skins are the product of the young of the Karakul or Arabi sheep and not of the Persian breed of sheep. These sheep are native of Bokhara, in Russian Turkestan, and are not found in Arabia, and only to a small extent in Persia. A number of other terms have been used in connection with the industry some of these being used interchangeably with persian lamb. Among these are Broadtails, Astrachan and Krimmer. The term "Broadtail" is applied to skins of lambs of Karakul blood and born before the close of the regular gestation period. Astrachan and Krimmer skins are supposed to come from sheep of somewhat different breeding.

23

Demand and Supply of Skins

The demand for Persian lamb skins has increased wonderfully during the past fifteen or twenty years and is still expanding. A member of the largest importing firm in America is of the opinion that there is no immediate indication that the supply will exceed the demand. The higher prices paid for skins has led to a great deal of crossing for the purpose of procuring a greater supply of skins, and it is held by some authorities that the very existence of the breed in Bokhara was threatened.

Volume of Trade

The skins imported to this country come over in the raw state in bales containing around 100 skins each. They are unsorted and some of them are not worth more than twenty-five cents each, but most of them range in value between $3.50 and $15.00. It has been estimated that $14,000,000.00 are spent abroad annually for skins and this may indeed be possible, for one New York house alone handles from 200,000 to 250,000 skins per season.

Dr. Young the only Importer

The possibility of establishing the industry in America led to two importations being made in the years 1908 and 1912, respectively. These sheep were brought over by Dr. C. C. Young of Belen, Texas. The first lot consisted of five rams and twelve ewes and the second of twelve rams and seven ewes. From this stock and its offspring, flocks have been established in Texas, New Mexico, Kansas, Maryland and Prince Edward Island, Canada.

Characteristics of the Breed

The Karakule is a hardy, broad-tailed, medium sized sheep of considerable length. The rump is characteristically rounded and usually steep. The rams are horned but the ewes are usually hornless. The ears are small and pendulous. The face is narrow and much rounded and together with the legs is covered with short, glossy hair. The body of the adult bears a coarse, long, hair-like wool, varying in color from light gray to black. The absence of soft under wool is said to be an indication of purity of blood. The mutton of the Karakul is said to be of very high quality.

Desirable Qualities in Skins

The lambs when dropped are usually a glossy black but rarely golden brown ones occur. The wool of the lamb is tightly curled over the body and well over the head and down over the legs. The qualities that determine the value of a skin are tightness and size of curl, the lustre, and size of the skin. The lustre is improved by the dyeing process which is essential in preparing the skin for use. The curls rapidly lose character and the lamb should be killed when not older than ten days, though there is much variation in the age at which the skins are of greater value.

The industry is still in its infancy in America and much is yet to be learned concerning it. Present indications point out a gradual progress and this is most desirable.

Fine wool crossing not Satisfactory

The Department of Agriculture in its work at the Experimental farm at Beltsville, Maryland, found that the Karakul cross upon the American Merino was unsatisfactory from a fur standpoint. Results from private flocks confirm this finding. This crossing has extended to include more of the breeds, and indications are that none of the close wool sheep give satisfactory results, especially in the first crosses. What can be developed from higher crosses containing a higher percentage of Karakul blood remains to be seen.

Breeding for Fur and Fecundity

The Karakule-Barbado cross was also tried at Beltsville. The Barbado is called the woolless sheep and the first cross resulted in a failure so far as curl was concerned, although the lustre was all that could be desired. In November 1913, the skins of eight lambs sired by a Karakul ram, out of first cross Karakul-Barbado ewes, were sent to New-York for valuation. One skin was appraised at fifty cents and one at $10.00. The average price of the eight skins was $4.75. The work is being continued and the higher Karakule crosses are being produced. If the high fecundity of the Barbado can be maintained in these crosses and the fur improved by continually using pure-bred Karakul sires, this may prove a means of increasing the amount of Karakul blood in America. Some Cotswold and Lincoln ewes are now being bred to a Karakul ram.

Barbado Fine wool Harmful

The method of removal and treatment of the lamb skin should be as follows: Cut a straight line down the belly, and also cut down on the inside of the legs to meet the center line. Do not cut off any part of the skin, leave on the ears, nose and tail to the tip. Be careful not to make unnecessary cuts. *Stretch* skin evenly on a board, fur side down and dry in a cool place. Do not salt the skin or double it up for shipment purposes. The principal object is to avoid cracking the skin. See that it is properly shaped when nailed down to the board and thoroughly dried before shipping. Do not sun dry the skin.

Skinning and Curing

Sales of Animals

The high price of breeding stock is at the present time a deterrent influence upon the industry. Such purebred rams as are available have sold at from $500.00 to $1,000.00 each. Ewes are somewhat cheaper. When buying breeding rams be careful to get pure bred animals. Some breeders claim that as good results can be obtained by the use of half blood stock, but this has not yet been established. It is advisable to buy only such rams as have already demonstrated their ability to sire skins of value.

No. 10. Karakul Ram, Teddy, Jr.—a son of Teddy, Sr. This ram was loaned to the United States Department of Agriculture and was killed by a kick from a zebra. His head may yet be seen in the office of Dr. George Rommel, Bureau of Animal Industry.

No. 11. Two coarse wool Karakul ewes with lambs at foot. The fur of these lambs is prime a few days after birth.

THE

CHAUTAUQUA MANAGERS ASSOCIATION

Suite 630, Orchestra Building,
220 South Michigan Avenue.
Charles W. Ferguson, President.
Alfred L. Flude, General Manager.

Chicago, Sept. 7, 1912.

Dr. C. C. YOUNG,
El Paso, Texas.

Dear DR. YOUNG:—

You know, of course, that we have in years gone by booked some of America's most famous orators, such as—

Senator BENJAMIN WHEELER
Ex-Governor JOSEPH W. FOLK,
Governor J. FRANK HANLEY,
Capt. RICHARD P. HOBSON and
Hon. ROBERT M. LA FOLLETTE.

Dr. Young a Defendant of Russia and that we are very particular as to whom we select when it comes to making addresses to our Chautauquas. We have followed your work with a great deal of interest the last few years, and have noticed that your articles, which have been many and which you have always written in defence of Russia, have always been well received, but we have never invited you to make addresses as we felt you were not sufficiently known.

In the last two or three years, however, you have received a great deal of mention in the press throughout the United States and we have desired to book you next summer if you feel you can spare the time. We feel that possibly only one side of Russia has been represented in the United States, and we would like to have you explain to our people the following points:—

1. When Russia was a Republic.
2. The Relations of the Emperor of Russia to the Greek Orthodox Catholic Church.
3. The Cossacks of Russia and their Relation to the Government.

4. The Jewish Problems.
5. Russia's Policy in Asia.
6. The causes that led to an entente between Russia and Japan.
7. The Connection of the Abrogation on our part of the Russian Treaty with the newly formed entente between Russia and Japan.
8. Russia's Passport System.

Awaiting your early reply, we are
Yours very truly,
CHAUTAUQUA MANAGERS ASSOCIATION,

(Signed) CHAS. W. FERGUSON,
President.

The following subjects were added later in place of Nos. 6 and 7.

Bokharan Fur and Rug Industries.
How to Raise Persian Lamb and Astrakan Fur in America.

The Karakúl Sheep Industry of America.

Mr. Joseph Simonson, Manager of the Middlewater Cattle
Company, Texas, tells the story of the industry
in America.

———

Reprinted from the Charlottetown Guardian, issue of
December 11th, 1913.

———

December 11th, 1913.

THE KARAKÚL SHEEP INDUSTRY

Mr. Joseph Simonson, President of the Middlewater
Cattle Co., and partner of ex-Congressman Loudon, son-in-
law of the late George Pullman of the Pullman Car Co.,
who has been in Charlottetown for the past few days, left
for Washington, D. C., Monday morning. Before leaving
he gave the following statement to The Guardian, more
in response to certain very direct questions that were
propounded to him by a number of Charlottetown's
most influential men at a small function held in his
honour at the residence of U. S. Consul Frost. Mr.
Simonson was greatly impressed with what he had seen
of Prince Edward Island and expressed the hope that
he would be able to spend his summers here with his
family who are now residing on the Loudon Ranch,
near Middlewater, Texas. **Mr.** Simonson is the only
man who stuck to Dr. Young for the first three years
when matters appeared somewhat against the Karakul
proposition. Following is his statement.

In June 1911 the Middlewater Cattle Company
of Middlewater, Texas, bought one half interest in Dr.
Young's first importation of Karakul sheep, which
by that time had increased from fifteen head brought
into the country in 1908, to fifty-two head. In the final
division our company got twenty-six head, the balance
remaining in the possession of Dr. C. C. Young.

What gave us especial confidence in Dr. Young's proposition was the fact that he refused to accept any cash whatsoever, preferring to take stock in the company for the entire amount that was due him, and of which stock, up to within a few weeks ago he never sold one share.

At the time of the division, Dr. Young stated that he might be obliged to sell his sheep, and for that reason all of the original five rams were taken by us, except **No Good Blood Disposed of**

No. 12. Mr. Joseph Simonson, Manager of the Middlewater Cattle Co., Texas. The ram is Teddy, Sr., as he appeared late in 1913.

one ram that had a great deal of fine-wool blood in him, and was known not to be satisfactory.

Among the four rams that became our property, **Kansas man Secured one Ram** there were two that showed no traces of finewool blood, and we had already the necessary proof that these two rams were by far the best rams, and especially was that true of Teddy Sr., the only ram that produced tight curled skins when bred to the full blood ewes that were satisfactory, especially so when bred to the two ewes that were better than the rest of them. To give Dr. Young's herd a chance to demonstrate what a good ram will produce, he was given a son of Teddy that was satisfactory, and which later died near Canutillo,

Texas. From this ram however there came one good ram-lamb which was later sold to the University of Edinburgh, and during Dr. Young's absence in Asia was removed from his El Paso ranch by one Crawford, who together with one McCombs, had purchased the entire herd from the Doctor, after being told that they had to procure good rams. These facts came to me through Dr. Young, whose correspondence with the Edinburgh people I had read. That the one good ram above mentioned died as well as the one Karakul fine-wool ram of the original importation of five I know to be a fact, and the skins are today in El Paso where they can be seen.

One source of Blood not enough

Again, shortly before Mr. Crawford purchased this herd, Dr. Young informed me of his great misfortune, so that there is absolutely no doubt that the little ram-lamb above mentioned, and which Dr. Young tells me he is now trying to get back through the courts, is the only lineal descendant of the only first class ram ('Teddy Sr.) of the first importation that one outside of ourselves could own. Assuming now that Dr. Young would have failed in his efforts of securing another importation, what chance would any herd of sheep have with one source of blood supply? Would any sane man risk his money on a flock of Karakul sheep that contain any amount of fine wool blood, which experience has shown us to be fatal to tight curl formation when possessing only one coarse wool ram with which to breed out this fine wool strain?

Crawford Herd are Finewools

One must not forget that the other rams in the Crawford herd are bound to be contaminated with fine-wool as they can only be descendants of three Karakul finewool rams of the first importation.

Fasset is Fairly Satisfactory

What about the second Karakul coarse wool ram heretofore mentioned, which as we stated already was proven to be free from fine wool blood? It happened so that before the division of the sheep that that ram was not bred to the full bloods, and when bred to Shropshires, quite a few of them came red, showing him to be too closely related to the red fatrump varieties of Central Asia, which results in a curl that has, (per se), enough luster, but rather open curls, something that is not desired, and for that reason I never bred him to the Karakul ewes. This same ram when bred to Lincoln ewes at Roswell, N. M. gave fairly satisfactory results, and for that reason I have never made any efforts to sell him.

In August 1912 Dr. Young informed me that he intended to bring from Asia two good unrelated coarse wool Karakul rams for the Crawford and McCombs sheep in order to breed their ewes up to the proper standard, as he intended to retain an interest in that flock. I made the same arrangement with him for the Middlewater Cattle Company, and was not a little surprised when the Doctor returned without the absolutely necessary two rams for the Crawford and McCombs sheep, telling me that he could not bring them, as Crawford failed to put up the money. When Professor Nabours of the Manhattan Agricultural College of Kansas visited our ranch in June 1913, he told me that Crawford's action was a colossal blunder. In view of this I am not a little surprised to find that the Secretary to the President of the same college has in the past year been writing numerous articles in several papers including "The Country Gentleman," in which he speaks of the wonderful success of the Crawford sheep, always mentioning Professor Nabours, never once indicating the real true facts regarding the most limited good blood supply of that flock.

When any one thinks that blood supply is not of the most vital importance in raising good animals, then let him deny that there is not danger in inbreeding.

When Dr. Young left for Asia, I, as president of our company, felt that he should not only bring in two unrelated rams for the company, but at least one half dozen. However the great expense of the animals and the fact that various tests proved unsatisfactory, due to our ignorance of the business in not knowing that our native fine wool ewes such as Merinos, Shropshires and others would give us no tight curled skins in the first cross, made it impossible for me to convince our principal stockholder, that the future of our industry depended entirely upon Dr. Young's success in getting more good rams into America.

When Dr. Young arrived with eleven rams in this country in April this year, I was hopeful that all of them would be brought to his ranches in El Paso and Roswell. I received my first shock when I learned from the Doctor that one ram was killed in quarantine at Baltimore, but was pleased that the remaining rams were sent to El Paso and Roswell, which is not far from our ranch at Middlewater.

Second Importation Fine Specimens

In June 1913 I selected our two rams from four rams that were on the Doctor's ranch at El Paso, and while there learned from him that they had the best chance of being absolutely unrelated, due to the fact that these rams were raised a great distance apart.

The natives of Central Asia without exception inbreed, and so do even those few Russians that have in the past few years managed to get a few sheep out of Bokhara.

I paid Dr. Young $2,500 cash and gave him two good coarse wool rams of our flock, and secured from him the promise that he would exchange rams with us, in order to prevent inbreeding for four generations, after which time I figured on exchanging with the doctor's associates at Roswell.

Mexican Revolution Causes Sale

On account of the Mexican trouble on the border where the doctor's ranch is located he sold one half interest in his sheep to parties in Charlottetown, Prince Edward Island, which worried myself and associates greatly, and never having had the unqualified support from some of our stockholders I was in hopes of consolidating our interests with the Doctor and associates. In this I failed and more to please the doctor than myself, I sold the entire herd to him, with the understanding that when he has consolidated all good sources of blood supply and formed a strong parent company, a subsidiary company shall be started at Middlewater, in which the doctor and his associates and myself could become interested. I am now considering this proposition which depends somewhat upon my ability to secure the right kind of grazing land in our country.

Getting $12 Skins

In reading the circular recently issued by the United States Department of Agriculture, I notice that they have had no satisfactory results in their first cross from such fine wool sheep as Merinos, Shropshires and Barbados, and the latter which Dr. Young tells me are also seriously contaminated. It has cost us a great deal of money and several years of experimenting to learn that $10 and $12 skins can only be produced from Karakul rams, free from fine wool where they are crossed to such coarse wool—long wool sheep as Lincolns, Cotswolds, and red Persian Fatrumps that are entirely free from fine wool admixture, and have the coarsest of wool.

The Best Native Ewes

I believe the Black Faced Highland with its coarse-wool when free from fine wool, will excel the Cotswolds and Lincolns. If one is satisfied with $2.50 and $3.00

skins, then one may use Karakul rams with fine wool blood in them, providing they are crossed to Lincolns or Cotswolds that have great lustre and very coarse wool.

If one takes into consideration that ewes whose lambs are killed the first few days after birth, may be bred again the same year, and thus assuring the breeder two crops within thirteen months, (this is true in the southwest) then one may be justified to buy inferior Karakul rams, but my advice to all breeders is to start with coarsewool Karakul rams and coarsewool long-wooled American ewes, that give excellent results in the first cross. It would appear that it would be for the best interest of the Karakul industry at large, if all the Karakul ewes now in America were bred up to the proper standard by the best rams.

Two Crops a Year

When I bought our first herd from Dr. Young, he exhibited a certificate from the Poltava Agricultural Society of Russia, to the effect that all the fifteen sheep of the original importation were of "plethoric" Kara-kule variety. The use of the word "plethoric" rather perplexed me, but its literal translation is "full blooded." One needs to look but once at the two good rams, and two good ewes of the original importation to notice the great difference between the good Karakul and those that are not desirable. Those that desire to satisfy themselves of that fact need but look once at the sheep, now on the ranch of Dr. S. R. Jenkins, where both varieties can be seen and watch the results next Spring when the lambs come.

Term "Full-Blooded" Misleading

Dr. Young tells me that he has issued a certificate to Crawford and McCombs to the effect that the Poltava Agricultural Society has given him a certificate, stating that all numbered and unnumbered sheep originally imported by him were full bloods of the Kara-kule varieties. Those promoters in Nova Scotia that are now exhibiting that certificate do not misrepresent it at all, but the future will tell the story, which, if correctly represented I am positive will tally closely with my statements, namely, that no skin really worth producing will ever come from a Karakul ram with fine wool blood in it, no matter to which of the three good classes he may belong.

Finewool Sheep Produce only Astrakhan Fur

AMERICAN EMBASSY,
St. Peterburg.

July, 21, 1908.

To the EDITOR OF COLLIER'S WEEKLY,
New York.

SIR:—

At the instance of Dr. C. C. Young, an American citizen of Russian origin, I take pleasure in stating that Dr. Young came to this Embassy some five months ago bearing letters of introduction to the Ambassador from President Roosevelt and Secretary Root. During the past five months he has travelled extensively through Russia and has been granted free transportation by the railway administration. The Embassy aided him in getting interviews with several Russian Ministers and other persons of note.

I am, Sir,

Very sincerely yours,

(Signed)   MONTGOMERY SCHUYLER, Jr.
Charge d'Affaires.

————

U. S. DEPARTMENT OF AGRICULTURE

BUREAU OF ANIMAL INDUSTRY,
Washington, D.C.

October 7, 1912.

DR. C. C. YOUNG,
care American Consulate,
Moscow, Russia.

My dear Doctor YOUNG,

In reply to your letter of September 17, I wish to say that as soon as I received this letter, I called up Mr. Knorr, and was informed that the American Breeders Magazine had been published and it was too late to make any corrections in your article.

I have taken the matter up with Doctor Hickman **Quarantine** of the Quarantine Division regarding the feeding of **at** your sheep in quarantine at Baltimore, Dr. Hickman **Baltimore** will write you.

I note what you say regarding the young ram that you expect to ship to Europe. If you do not ship this ram, as intended, it is very likely that we could use him to good advantage, owing to the fact that we have a number of ewes that should be bred this fall.

Regarding the letter written to Mr. Rommel by Mr. Herman Basch, we can do no better than to send you a photostat copy of the same.

Trusting that you are enjoying yourself, and that you will be able to make a nice importation of Karakul.

I am,

Very truly yours,

E. L. SHAW,
Senior Animal Husbandman in
Sheep and Goat Investigations.

Letters of Herman Basch & Co., fur dyers, to Dr. George Rommel of the United States Department of Agriculture.

———

NEW YORK CITY.
18 West 27th St.

Sept. 11, 1912.

DR. GEORGE ROMMEL,
Washington, D.C.

DEAR SIR,

Dr. Young has just brought to me 6 skins which some time ago I tanned and dyed for him. Mr. Speer and myself have priced them and the Doctor desires that I should quote to you the *Real Wholesale Market Price*, this is, in ball lots of 160-200 skins.

Prices of Skins

Skins No. 12 and No. 9 were priced by Mr. Speer before he left for Russia,—skin No. 9 being valued at $10.00 and skin No. 12 at $8.00. The other skins I priced myself: No. 5 $9.00, No. 4 $7.00, No. 29 $6.00 No. 32 $6.00. These skins compare favourably indeed with those which are annually imported from Russia.

Very truly yours

(Signed) HERMAN BASCH.

Photo by Dr. C. C. Young.

No. 13. A four weeks old Karakul lamb held by a neice of Dr. C. C. Young in Texas.

Will it pay to raise Karakúl Sheep?

(By Dr. C. C. Young.)

**How to
Breed
Persian
Lamb Fur**

When a Karakul free from fine wool is bred to the common English longwool breeds of sheep, selected for coarseness of wool fibre and freedom from any fine admixture of underwool, the lamb or lambs resulting from this cross, are black, lustrous, curly, individuals whose skins if taken at their primest condition a few days after birth, are worth from five to thirteen dollars in wholesale lots.

However, it would be unwise owing to the scarcity of good Karakul blood to slaughter the above-mentioned half-blood lambs, because the great majority of them, if crossed with the same type of common sheep, will produce the same valuable persian lamb fur in the quarter-blood Karakul offspring.

Thus "persian lamb", as the skin of the young of the Karakul sheep is named in commerce, is often produced from lambs which have not more than twenty-five per cent of Karakul blood.

Even the quarter blood Karakul sheep if bred to the same type and breed of long wool will frequently produce the Krimmer or the Astrakan fur in the one-eighth blood lambs.

**Demand for
Breeding
Animals**

Thus, only a few available blood lines of Karakul blood represented by several unrelated rams, which would prevent inbreeding, and which are free from fine wool of the Merino or Shropshire type, is necessary in order to produce the highest grade of persian lamb and broadtail fur from our common flocks of Lincolns, Cotswolds and Highland Black-face ewes. Owing to the great popular interest, particularly in Canada, in the production of fur under domestic conditions, a great demand for Karakul rams has sprung up. There are none for sale at the present time, except some fine wool Karakuls which I parted with while the industry was yet in an experimental stage in Texas. Breeders will be sorely mistaken if these fine-wool Karakul rams are purchased and bred to their longwool ewes. The resulting offspring will be black, but the fur will only rarely grade higher than "Astrakan" and be worth but from four dollars at the best to skins worth not more than one dollar. Fine wool destroys the curl and lustre and gives only moderate profits.

All Black Lambs not Persian Lamb

The popular demand for Karakul rams is well illustrated in a letter recently received from Professor Robert K. Nabours, head of his department in the Kansas State Agricultural College, Manhattan, Kansas. He said: (November 18, 1913).

"I did not learn whether or not you have any pure-bred Karakul rams for sale. This is the one question everybody asks: 'Where can I secure a pure-bred Karakul ram and what is the cost?' It must be that I have five hundred or more such inquiries on file in my office and I have not been able to give a satisfactory answer. I wish you or somebody else could bring over a thousand Karakul rams. They would go like 'hot cakes'. There is more interest in Karakul sheep now throughout the United States and Canada than in any other form of live stock".

Professor Nabour's Statement

I have repeatedly refused to sell good full-blood rams at $2,500.00 each as I had not developed the parent company to the point where rigid selection of breeding stock, and control of the blood lines was assured until this year,—1914. During the summer of 1914 a small number of full-blood bucks could be absent from the parent flock on certain conditions. and considerable number of half-blood bucks will be available as soon as tested in the breeding pens. These will especially be recommended for fur production on sheep ranges and will be as valuable for that purpose, in most cases, as the full-blood bucks. When it is desired to breed pedigreed, full-blood stock it will be necessary to write to the parent company at Charlottetown to ascertain the conditions on which full-blooded stock will be sold. Proved half-blood bucks should sell in any quantity for many years at several hundred dollars each and full-bloods at several thousand dollars each.

Proposed Disposal of Stock

Inasmuch as one full-blood ram and one hundred common ewes will produce about fifty half-blood bucks each year, of which the great majority should prove to be valuable fur getters, the huge profits of the Dr. C. C. Young Karakul Sheep Company, Limited, can be appreciated. Rams can easily be mated to 100 ewes each by the method known to shepherds as "hand breeding".

Profits in Sale of Rams

No. 14. Two coarse wool Karakul ewes and two eight months old ewe lambs on the author's farm—Bunbury—near Charlottetown, P. E. I., Canada.

PHOTO BY Dr. C. C. YOUNG.

No. 15. Karakul sheep on the estate of Jakoulav, Province of Ufa, on the Asiatic border. Dr. Young has inspected all the European herds and the Asiatic herds wherever his former passports permitted him to go.

42

It is the polygamous characteristics of the ram, **Demand for** combined with the facts that sheep are fed inexpen- **Persian** sively on grass, and are truly domesticated animals, **Lamb** that has given the industry of persian lamb fur production its great advantage over silver fox fur production. The demand for persian lamb fur is also fifty times in excess of the demand for silver fox and, in America, is one-third as great as the demand for all other furs combined. It is a durable, warm, light, fur, worn by both men and women and is bound to increase in demand. We have data enough on hand to assure ourselves that our half-blood rams will produce skins worth an average of from six to ten dollars each, wholesale. Probably the average price will be raised yearly by selection as well as by the gradual increase in the price of fur generally. Persian lamb fur has increased in price 300 per cent in twenty years.

Taking the average value of the skins at seven **Breeding** dollars each huge profits to the fur raiser on stock ranges **for Fur** are assured. Ordinary range lambs at six months old, fetch about $3.00 each and Karakul lambs at 2 days old fetch $7.00 and upwards. Besides, if the lamb is born prematurely its skin is yet more valuable, or if it is born dead or dies soon after birth, as from ten to thirty per cent do, the skin is still valuable. We may look for the general establishment of Persian Lamb, Broadtail, and Astrakhan production on farms as well as on ranges, while **Breeding** the sale of breeding stock will be a highly profitable **for** business for those who can get possession of the blood **Pedigreed** lines and develop a fancy pedigreed herd by selection. **Stock**

One Karakul ram and 100 Lincoln ewes will produce 80 half-bloods at a conservative estimate.

COST OF BREEDING STOCK

1 Karakul ram	$2,500.00	**Profits for**
100 Lincoln ewes (selected for fleece)	1,000.00	**First Year**
Cost of feed and attendance	500.00	
	$4,000.00	
40 half-blood rams at $250.00 %	10,000.00	
Total receipts	$10,000.00	
Deduct cost	4,000.00	
1st year's net profit, 150 per cent, or	$6,000.00	

Improvement in weight and Mutton Quality

The owner of sheep ranges stocked with fine wool sheep like Shropshires, the Down breeds, Merinos, Romboulets, etc., may take comfort in the fact that persian lamb fur can be produced in two crosses on their fine wool stock. The first cross gives an average of thirty per cent increase in weight, and greatly improves the mutton —according to tests made by Armour and Company—and gives black lambs of wonderful hardiness and rustling

No. 7. Karakul Ram lamb one month old.
His mother is a medium fine wool Karakul and his father a coarse wool Karakul. The curls are becoming open and the fur is therefore becoming less valuable daily.

qualities. A considerable number of the skins are good Astrakhan fur. In the second cross ten dollar skins are frequently produced providing a coarse wool Karakul ram, free from fine wool blood, is used. It should be noted that the herds of Bokhara are largely composed of Afghan fine wools crossed in some instances with Persian fat rumps, while pure coarse wool Karakul representatives are very few in number. Owing to the lack of breeding intelligence among these people, the Karakul strain is not kept pure and the production of persian lamb fur will soon cease in that country. Furriers can testify to the decline in quality in recent years owing to fine wool introduction. This is an additional reason for haste in establishing the industry in America.

Bokharan Persian Lamb Deteriorating in Quality

The White House.
Washington.

Jan. 29, 1914.

My dear Senator Sheppard,

In compliance with your request, I have been glad to write the charge d'Affaires at St. Petersburg, asking him to extend to Dr. C. C. Young every proper courtesy.

Sincerely yours

(Signed) J. P. Tumulty,
Secretary to the President.

Hon. Morris Sheppard
United States Senate.

———

THE FIRST KARAKÚL LAMB.
Born in Canada.

———

The first Karakul lamb born in Canada was dropped on Upton Farm, the property of Dr. S. R. Jenkins, Charlottetown, on February fifth, 1914. It was tightly curled even on the belly and legs, was glossy and lustrous, of a good size and healthy. His life was spared him past the valuable fur stage and he will grow up gradually losing his jet black colour and oxidizing to a grey like his parents. It is the property of the Dr. C. C. Young Karakul Sheep Co., Limited of Charlottetown.

———

POINTS FOR FARMERS

TO

CAREFULLY EXAMINE AND CONSIDER.

1. Karakul sheep are domestic animals,—and the problems of domestication have been settled for ages.

2. Sheep live on grass—not on costly meat and milk like foxes, mink, marten and otter.

3. One male will mate with 100 females, if the hand breeding method is employed. Hand breeding means simply the immediate removal of the ewe from the breeding pen after mating.

4. The lambs have not to be reared to maturity for fur. Lambs born dead, or whose mother dies before the regular period of gestation is ended, are more valuable for fur than any others.

5. Only 25 per cent of coarse wool Karakul injected into such breeds as Lincolns, Cotswolds, Highland Black-faced, and other coarse wool sheep, which are free from fine wool admixture, will produce lambs whose skins are "Persian lamb" fur.

6. The volume of wholesale trade in Persians skins for the United States is $14,000,000.00 or greater in amount by over twenty times than the total trade for the world in Silver Fox skins.

7. The meat of the young lamb is declared to be a great delicacy by Professor Robert Wallace of Edinburgh University and is highly prized by Asiatic gastronomists.

8. Persian lamb is a staple fur—there can be no overproduction.

9. Persian lamb skins fetch $10.00 each wholesale, while 6 months old lambs of native sheep bring four dollars each.

10. Is there not a great future in America for Karakul sheep? Will not the sale of full-blood and half-blood rams for breeding purposes be a profitable industry for many years to come? Will not the production of Persian lamb fur from grade Lincoln or Cotswold × Karakul crosses be a most profitable industry on stock ranges?